THE SMALL MINERS HELPER

I0489895

This short book was written to assist the small miner by presenting some of my own experience in mining and processing. The processes are in some cases past down by former colleagues, Old timers from the depression era, my own hard work experience, some internet facts, formal education, and opinion. Most of all it was written for the purpose of passing down some answers and to keep the interest going for the small miner.

Each Chapter (Very short) touches on different aspects of the mining industry, but also leaves room to continue discussion in another short book, regarding that subject which will be published at a later date.

Table of Contents

- Melting
- Smelting
- Basic Smelting using Copper as a collector
- Basic Smelting using Lead as a collector

Chapter 6: Assay by Fire (Fire Assaying)

- Basic Instruction

Chapter 7: Selling Precious Metals and Dore'

- What is Dore'
- What's my Dore' worth
- Knowing about Assays
- Choosing a Refiner

THE SMALL MINERS HELPER
A Basic overview of Mining and Processing in the US

Chapter 1

'The Miner'

A Nostalgic look at a miner removing concentrate and gold from a sluice box.

Successful mining has gone on for centuries, mostly the trial and error method of locating and digging up the gold in placer areas is the most common scenes known today as mining for gold.

Truth is much more of the 'Hardrock' mining is where people really made their fortunes, leading them to large pockets of the yellow stuff along with massive amounts of the monitory metal silver.

Mining has been done dating back to the ancient ages of man, seeking out such things as gold, silver, and colorful gems for trade and personal adornment. In the 1820s the southeastern US had big strikes the areas of Georgia and Alabama. Many Miners flocked to the lawless California during the 1840s to seek their fortunes only to find no gold but poverty and death awaiting them.

There is not a state in the US where given the right conditions, a fortune cannot be made. Gold and Silver still remain abundant in the seven western states where much of the land belongs to cattle barons and the US government.

Miners come in all shapes, sizes and denomination of race and gender. Some strong, some weak, some Intelligent, and some well....All seeking the big strike!

The truth is today, the small recreational prospector 'Weekender' can expect to spend upwards of $5000.00 before finding their first Cache of $100.00 worth of gold or silver, spend relentless hours on research, and tiresome backbreaking work. You ask do I think it's worth it, you bet it is because 1 in 10,000 has the potential to 'strike it rich'

A small group of independent miners can expect to pay up to $50,000.00 for equipment, (Rental or purchase used and repair), $5000.00 in Lab fees, and $20.000.00 in government permits, lawyers, and consultants in just the exploration phase of their mining operations. However the money spent usually pays off in a very short period of time, especially since most of these expenses are tax deductible and exploration saves huge amounts of wasted time and money.

Miners both small and large will always continue their quest for the precious metals, because if one thing is for sure, history has proven the profits of the venture.

Chapter 1

'Finance'

When it comes to venturing of mining precious metals for profit the first thing that comes to mind is, how do I pay for this?

Well if you are a little guy, don't quit your day job! Having funds available to pay for small equipment, testing and prospecting tools, can be handled with a small dip into salary or savings. But remember unless you are 1 in the million out there you will be running on luck, just to say spend it and it will come. Nothing is farther from the truth.

Even the little guy needs to plan, consider costs or add thing to what you see below.

1. $ for Travel
2. $ for special clothing
3. $ for prospecting gear
4. $ for testing (Assays, pilot ops)
5. $ for medical first aid and safety (Important)
6. $ for permits (land management entities etc.)
7. $ for Legal advise
8. $ for mining equipment
9. $ for processing equipment
10. $ for anything that you can imagine to come up.

Nothing is free not even the minerals in the ground, everything has a price tag attached to it. The above is referred to as 'putting the horse before the cart'.

If the finance is in place total concentration can be afforded to optimizing your efforts. A lot of miners fall flat on their face because the first plan they make is not FINANCE. They then sell their equipment for nothing and generally profess that it just simply could not be done.

THE SMALL MINERS HELPER
A Basic overview of Mining and Processing in the US

There are many options for finance, credit report dot com is not one. If you don't have the money, don't spend it.

Very small and Junior Mining companies usually get private investors to front the money for startup and operation however a business plan with theoretical timelines on accomplishing the completion of a stage of operation is a must. Since they will be gambling with your money, it is important for you to know the stakes are in their corner (NOT YOURS) to win. Remember accountability and performance prevails.

The same holds true of the juniors when they turn to Public Ventures as accountability is much higher and takes professional people to put it all together. Junior mining companies seem to always profess and flaunt their reserves and holdings but the real proof is which stage of the game their in. Are they actually going to produce? Or are they two retired geologists with a large portfolio sitting in a plush office somewhere drawing a large salary from the investors? Are they even out of the pre-feasibility stage of exploration? One must look at exactly what they are achieving along with expected real timelines to determine if this is a venture which is going to produce or just another scam. Things such as forward looking statements with legal declaimers are generally just words to confuse the public layman.

Working for several of these outfits gives one a real look at what really goes on. Everything with the word mining attached to it is extremely expensive. From the equipment to the expensive professionals to the miners and processors which perform the labor, and not to forget the top of the building, where founders CEO's and Top management draw huge salaries to promote a business which might not even be productive.

E for effort build these businesses, see what is real, can they produce, some of these outfits actually get things done, still very few will ever cover the costs invested (More commonly referred to as "Selling the Dream").

Large mining companies as seen today on the stock exchanges worldwide owe their very existence to investors and shareholders. Generally they are huge land barons and enormous financial machines. Mining itself plays a part in the mix but the financial machine is everything.

Remember when considering being a miner or mining operation the key to success is financing it all without the miner turning into a scam artist.

Chapter 1

'Small Mining Companies'

Nothing is more precious than seeing the small individual or small mining company strive for success. It seems it is usually 'one man's dream' which evolves into a profitable working machine.

Whether you're a one man venture, Mom and Pop or Junior venture the following notes apply to you.

Safety and training is of the primary and first concern. If you don't know or have not the expertise when dealing with mining material, the equipment, or elemental hazards, then go no further until you obtain it. Mining like many other industries has very specific hazards and procedures.

Training with specific equipment can be obtained by;

1. Experienced Miners
2. Equipment manufacturers
3. Testing Laboratories
4. Mining engineers

If you cannot find answers they in most cases can direct you to someone who has the experience to teach or answer any question you may have. Inexperience only leads to failure and hazardous conditions.

Safety on the other hand is mostly common sense, but in the case of most of us taken for granted which can put us in bad situations. Some things to remember before you start to mine.

1. Know your intentions of what you plan to do before starting.
2. Consider your methods
3. Know the area in which you plan to work, (i.e. How far it is to medical help, do I have phone service, what will happen if this situation arises). How can I get help in an emergency?
   Terrain, isolation, vehicle design, phone or radio capabilities all play an important part in a successful venture.

Many of the questions concerning safety in mining can be obtained in private by just going to the web and MSHA's web site. Listed here; http://www.msha.gov/   just go to their Education and Training section and get what you need, it's free.

Please note that all who involve themselves in mining should have the proper training before attempting to mine, haul, mill, or process minerals. It's only common sense!

Very small ventures such as an individual or 2 to multiple person ventures may be exempt from some governmental stipulations and requirements. However this does not mean that you are exempt from Federal, local, and state laws. Be sure to research this well, a mistake could lead to losing your equipment, property, and right to mine as well as some heavy fines.

Please note that MSHA for instance can fine you for any reason at any time on a mine site and that the inspectors interpretations of the law, are considered law. They can order closure at anytime and are not there to point out safety hazards as such agencies as OSHA but to fine you for any reason they can find.

Know well the 1872 Mining laws here is the web page provided by the BLM to acquaint you with some basics.
http://www.blm.gov/pgdata/etc/medialib/blm/ak/aktest/minerals/minerals_pdfs.Par.12787.File.dat/43cfr3715.pdf

Below is a Summary of the actual 1872 Mining Law

To view this summary see Volume 30 of the United States Code of Federal Regulations for the complete 1872 Mining Law

§21. Mineral lands reserved

In all cases lands valuable for minerals shall be reserved from sale, except as otherwise expressly directed by law.

§22. Lands open to purchase by citizens

Except as otherwise provided, all valuable mineral deposits in lands belonging to the United States, both surveyed and unsurveyed, shall be free and open to exploration and purchase, and the lands in which they are found to occupation and purchase, by citizens of the United States and those who have declared their intention to become such, under regulations prescribed by law, and according to the local customs or rules of miners in the several mining districts, so far as the same are applicable and not inconsistent with the laws of the United States.

§23. Length of claims on veins or lodes

Mining-claims upon veins or lodes of quartz or other rock in place bearing gold, silver, cinnabar, lead, tin, copper, or other valuable deposits, located prior to May 10, 1872, shall be governed as to length along the vein or lode by the customs, regulations, and laws in force at the date of their location. A mining-claim located after the 10th day of May 1872, whether located by one or more persons, may equal, but shall not exceed, one thousand five hundred feet in length along the vein or lode; but no location of a mining claim shall be made until the discovery of the vein or lode within the limits of the claim located. No claim shall extend more than three hundred feet on each side of the middle of the vein at the surface, except where adverse rights existing on the 10th day of May 1872 render such limitation necessary. The end lines of each claim shall be parallel to each other.

§26. Locators' rights of possession and enjoyment

The locators of all mining locations made on any mineral vein, lode, or ledge, situated on the public domain, their heirs and assigns, where no adverse claim existed on the 10th day of May 1872 so long as they comply with the laws of the United States, and with State, territorial, and local regulations not in conflict with the laws of the United States governing their possessory title, shall have the exclusive right of possession and enjoyment of all the surface included within the lines of their locations, and of all veins, lodes, and ledges throughout their entire depth, the top or apex of which lies inside of such surface-lines extended downward vertically, although such veins, lodes, or ledges may so far depart from a perpendicular in their course downward as to extend outside the vertical side-lines of such surface locations. But their right of possession to such outside parts of such veins or ledges shall be confined to such portions thereof as lie between vertical planes drawn downward as above described, through the end lines of their locations, so continued in their own direction that such planes will intersect such exterior parts of such veins or ledges. Nothing in this section shall authorize the locator or possessor of a vein or lode which extends in its downward course beyond the vertical lines of his claim to enter upon the surface of a claim owned or possessed by another.

§27. Mining tunnels; right to possession of veins on line with; abandonment of right

Where a tunnel is run for the development of a vein or lode, or for the discovery of mines, the owners of such tunnel shall have the right of possession of all veins or lodes within three thousand feet from the face of such tunnel on the line thereof, not previously known to exist, discovered in such tunnel, to the same extent as if discovered form the surface; and locations on the line of such tunnel of veins or lodes not appearing on the surface, made by other parties after the commencement of the tunnel, and while the same is being prosecuted with reasonable diligence, shall be invalid; but failure to prosecute the work on the tunnel for six months shall be considered as an abandonment of the right to all undiscovered veins on the line of such tunnel.

§28. Mining district regulations by miners: location, recordation, and amount of work; marking of location on ground; records; annual labor or improvements on claims pending issue of patent; co-owner's succession in interest upon delinquency in contributing proportion of expenditures; tunnel as lode expenditure

The miners of each mining district may make regulations not in conflict with the laws of the United States, or with the laws of the State or Territory in which the district is situated, governing the location, manner of recording, amount of work necessary to hold possession of a mining claim, subject to the following requirements: The location must be distinctly marked on the ground so that its boundaries can be readily traced. All records of mining claims made after May 10, 1872, shall contain the name or names of the locators, the date of the location, and such a description of the claim or claims located by reference to some natural object or permanent monument as will identify the claim. On each claim located after the 10th day of May 1872, and until a patent has been issued therefor, not less than $100 worth of labor shall be performed or improvements made during each year. On all claims located prior to the 10th day of May 1872, $10 worth of labor shall be performed or improvements made each year, for each one hundred feet in Length along the vein until a patent has been issued therefore; but where such claims are held in common, such expenditure may by made upon any one claim; and upon a failure to comply with these conditions, the claim or mine upon which such failure occurred shall be open to relocation in the same manner as if no location of the same had ever been made, provided that the original locators, their heirs, assigns, or legal representatives, have not resumed work upon the claim after failure and before such location. Upon the failure of any one of several co owners to contribute his proportion of the expenditures required hereby, the co owners who have performed the labor or made the improvements may, at the expiration of the year, give such delinquent co-owner personal notice in writing or notice by publication in the newspaper published nearest the claim, for at least once a week for ninety days, and if at the expiration of ninety days after such notice in writing or by publication such delinquent should fail or refuse to contribute his proportion of the expenditure required by this section, his interest in the claim shall become the proper6ty of his co-owners who have made the required expenditures. The period within which the work required to be done annually on all unpatented mineral claims located since May 10, 1872, including such claims in the Territory of Alaska, shall commence at 12 o'clock meridian on the 1st day of September seceding the date of location of such claim.

Where a person or company has or may run a tunnel of the purposes of developing a lode or lodes. Owned by said person or company, the money so expended in said tunnel shall be taken and considered as expended on said lode or lodes, whether located prior to or since May 10, 1872; and such person or company shall not be required to perform work in the surface of said lode or lodes in order to hold the same as required by this section. On all such valid claims the annual period ending December 31, 1921, shall continue to 12 o'clock meridian July 1, 1992.

§29. Patents; procurement procedure; filling; application under oath, plat and field notes, notices, and affidavits; posting plat and notice on claim; publication and posting notice in office; certificate; adverse claims; payment per acre; objections; nonresident claimant's agent for execution of application and affidavits

A patent for any land claimed and located for valuable deposits may be obtained in the following manner: Any person, association, or corporation authorized to locate a claim under sections 21, 22 to 24, 26 to20, 29, 30, 33 to 48, 50 to 52, 71 to 76 of this title and section 661 of Title 43, having claimed and located a piece of land for such purposes, who has, or have, compiled with the terms of sections 21, 22 to24, 26 to 28, 29, 30, 33 to 48, 50 to 52, 71 to 76 of this title and section 661 of Title 43, may file in the proper land office an application for a patent, under oath, showing such compliance, together with a plat and field notes of the claim or claims in common, made by or under the direction of the Director of the Bureau of Land Management, showing accurately the boundaries of the claim or claims, which shall be distinctly marked by monuments on the ground, and shall post a copy of such plat, together with a notice of such application for a patent, in a conspicuous place on the land embraced in such plat previous to the filing of the application for a patent, and shall file an affidavit of at least two persons that such notice has been duly posted, and shall file a copy of the notice in such land office, and shall thereupon by entitled to a patent for the land, in the manner following: The register of the land office, upon the filing of such application, plat, field notes, notices, and affidavits, shall publish a notice that such application has been made, for the period of sixty days, in a newspaper to be by him designated as published nearest to such claim; and he shall also post such notice in his office for the same period. The claimant at the time of filing this application, or at any time thereafter, within the sixty days of publication, shall file with the register a certificate of the Director of the Bureau of Land Management that $500 worth of labor has been expended or improvements made upon the claim by himself or grantors; that the plat is correct, with such further description by such reference to natural objects or permanent monuments as shall identify the claim, and furnish an accurate description, to be incorporated in the patent. At the expiration of the sixty days of publication the claimant shall file his affidavit, showing that the plat and notice have been posted in a conspicuous place on the claim during such period of publication. If no adverse claim shall have been filed with the register of the proper land office at the expiration of the sixty days of publication, it shall be assumed that the applicant is entitled to a patent, upon the payment to the proper officer of $5 per acre, and that no adverse claim exists; and there after no objection from third parties to the issuance of a patent shall be heard, except it be shown that the applicant has failed to comply with the terms of sections 21, 22 to 24, 26 to20, 29, 30, 33 to 48, 50 to 52, 71 to 76 of this title and section 661 of Title 43. Where the claimant for a patent is not a resident of or within the land district wherein the vein, lode, ledge, or deposit sought to be patented is located, the application for patent and the affidavits required to be made in the section by the claimant for such patent may be made by his, her, or its authorized agent, where said agent is conversant with the facts sought to be established by said affidavits.

§30. Adverse claims; oath of claimants; requisites; waiver; stay of land office proceedings; judicial determination of right of possession; successful claimants' filing of judgement roll, certificate of labor, and description of claim in land office, and acreage and fee payments; issuance of patents for entire or partial claims upon certification of land office proceedings and judgment roll; alienation of patent title

Where an adverse claim is filed during the period of publication, it shall be upon oath of the person or persons making the same, and shall show the nature, boundaries and extent of such adverse claim, and all proceedings, except the publication of notice and making and filing of the affidavit thereof, shall be stayed until the controversy shall have been settled or decided by a court of competent jurisdiction, or the adverse claim waived. It shall be the duty of the adverse claimant, within thirty days after filing his claim, to commence proceedings in a court of competent jurisdiction, to determine the question of the right of possession, and prosecute the same with reasonable diligence to final judgment; and a failure so to do shall be a waiver of his adverse claim. After such judgment shall have been rendered the party entitled to the possession of the claim, or any portion thereof, may, without giving further notice, file a certified copy of the judgment roll with the register of the land office, together with the certificate of the Director of the Bureau of Land Management that the requisite amount of labor has been expended or improvements made thereon, and the description required in other cases, and shall pay to the register $5 per acre for his claim, together with the proper fees, whereupon the whole proceedings and the judgment roll shall be certified by the register to the Director of the Bureau of Land Management, and a patent shall issue thereon for the claim, or such portion thereof as the applicant shall appear, from the decision of the court, to rightly possess. If it appears from the decision of the court that several parties are entitled to separate and different portions of the claim, each party may pay for his portion of the claim, with the proper fees, and file the certificate and description by the Director of the Bureau of Land Management whereupon the register shall certify the proceedings and judgment roll to the Director of the Bureau of Land Management, as in the proceedings case, and patents shall issue to the several parties according to their respective rights. Nothing herein contained shall be construed to prevent the alienation of the title conveyed by a patent for a mining claim to any person whatever.

§35. Placer claims; entry and proceedings for patent under provisions applicable to vein or lode claims; conforming entry to legal subdivisions and surveys; limitation of claims; homestead entry of segregated agricultural land

Claims usually called "placers," including all forms of deposit, excepting veins of quartz, or other rock in place, shall be subject to entry and patent, under like circumstances and conditions, and upon similar proceedings, as are provided for vein or lode claims; but where the lands have been previously surveyed by the United States, the entry in its exterior limits shall conform to the legal subdivisions of the public lands. And where placer claims are upon surveyed lands, and conform to legal subdivisions, no further survey of plat shall be required, and all placer-mining claims located after the 10th day of May, 1872, shall conform as near as practicable with the United States system of public-land surveys, and the rectangular subdivisions of such surveys, and not such location shall include more than twenty acres for each individual claimant; but where placer claims cannot be conformed to legal subdivisions, survey and plat shall be made as on unsurveyed lands; and where by the segregation of mineral land in any legal subdivision a quantity of agricultural land less than forty acres remains, such fractional portion, of agricultural land may be entered by any party qualified by law, for homestead purposes.

§36. Subdivisions of 10- acre tracts; maximum of placer locations; homestead claims of agricultural lands; sale of improvements

Legal subdivisions of forty acres may be subdivided into ten-acre tracts; and two or more persons, or associations of persons, having contiguous claims of any size, although such claims may be less than ten acres each, may make joint entry thereof; but no location of placer claim, made after the 9th day of July, 1870, shall exceed one hundred and sixty acres for anyone person or association of persons, which location shall conform to the United States surveys; and nothing in this section contained shall defeat or impair any bona fide homestead claim upon agricultural lands, or authorize the sale of the improvements of any bona fide settler to any purchases.

§37. Proceedings for patent where boundaries contain vein or lode; application; statement including vein or lode; issuance of patent: acreage payments for vein or lode and placer claims; costs of proceedings; knowledge affecting construction of application and scope of patent

Where the same person, association, or corporation is in possession of a placer claim, and also a vein or lode included within the boundaries thereof, application shall be made for a patent for the placer claim, with the statement that it includes such vein or lode, and in such case a patent shall issue for the placer claim, subject to the provisions of sections 21, 22 to 24, 26 to20, 29, 30, 33 to 48, 50 to 52, 71 to 76 of this title and section 661 of Title 43, including such vein or lode, upon the payment of $5 per acre for such vein or lode claim, and twenty-five feet of surface on each side thereof. The remainder of the placer claim, or any placer claim not embracing any vein or lode claim, shall be paid for at the rate of $2.50 per acre, together with all the costs of proceedings; and where a vein or lode, such as described in section 23 of this title, is known to exist within the boundaries of a placer claim, an application for a patent for such placer claim which does not include an application for the vein or lode claim shall be construed as a conclusive declaration that the claimant of the placer claim has no right of possession of the vein or lode claim; but where the existence of a vein or lode in a placer claim is not known, a patent for the placer claim shall convey all valuable mineral and other deposits within the boundaries thereof.

§42. Patents for non mineral lands; application, survey, notice, acreage limitation, payment

(a) Vein or lode and mill site owners eligible
Where non mineral land not contiguous to the vein or lode is used or occupied by the proprietor of such vein or lode for mining or milling purposes, such nonadjacent surface ground may be embraced and included in an application for a patent for such vein or lode, and the same may be patented therewith, subject to the same preliminary requirements as to survey and notice as are applicable to veins or lodes; but no location made on and after May 10, 1872 of such nonadjacent land shall exceed five acres, and payment for the same must be made at the same rate as fixed by sections 21, 22 to 24, 26 to20, 29, 30, 33 to 48, 50 to 52, 71 to 76 of this title and section 661 of Title 43 for the superficies of the lode. The owner of a quartz mill or reduction works, not owning a mine in connection therewith, may also receive a patent for his mill site, as provided in this section.

(b) Placer claim owners eligible
Where non mineral land is needed by the proprietor of a placer claim for mining, milling, processing, beneficiation, or other operations in connection with such claim, and is used or occupied by the proprietor for such purposes, such land may be included in an application for a patent for such claim, and may be patented therewith subject to the same requirements as to survey and notice as are applicable to placers. No location made of such nonmineral land shall exceed five acres and payment for the same shall be made at the rate applicable to placer claims which do not include a vein or lode.

§43. Conditions of sale by local legislature

As a condition of sale, in the absence of necessary legislation by Congress the local legislature of any State or Territory may provide rules for working mines, involving easements, drainage, and other necessary means to their complete development; and those conditions shall be fully expressed in the patent.

However there have been several amendments to the law and to be sure an individual should contact a mining lawyer to correctly interpret the actual law as it stands today.

The Small miner should be well aware of their rights and be able to recite and prove the written code to any official. Most small miners may not be capable of spending the tens of thousands that a large entity can afford, enough said.

Small mining operations should also be familiar with State and County requirements to have a compliable problem free operation. Be sure to check with a reliable source before making an imposition.

A small amount of time checking with the jurisdictional authorities will save a large amount of time in the long run.

Chapter 1

'Large Mining Companies'

Large Mining companies in themselves are huge machines run by large corporations with all the access to Legal, Financial, and expert information needed for success of a company.

All aspects of these companies are comprised of experts from the Real estate, Labor, Scientific, Engineering, Financial and Legal disciplines. Unlike the Small Mining companies, the big boys are set up to handle just about anything and good at it too!

Planning and intentions of an operation may be made as much as 30 years in advance, unlike many businesses the game plan is in place and they simply do not fail.

Most of the large companies are placed by financial experts, Legal experts, and enormous investors, and listed on the stock exchange for public opportunity. There simply is 'no one guy' but many go to personnel.

These Financial giants also drive the market supplying not only the precious metals we seek but many industrial minerals as well.

They are the ones who face most of the public scrutiny, deal with most of the legal, economical and environmental issues in mining. It is thru them that the availability of a small opportunity in mining is available today. I'm sure without them the governments would have already banned mining by the small operation today.

The large mining companies also face large scrutiny by the public, however the ladies and men love their silver and gold and we all love our electronics, without the large operation the demand for precious metal simply could not be met.

These companies GENERATE a large amount of jobs for the American people from the mines themselves to the billions spent in support of these operations from education to equipment, all generated by this giant machine. It's great to say 'Not in my back yard' but without mining in this proportion the US would be broke today. Remember mines produce real value, unlike the printers which produce pretty paper, "think we call them dollars". It's up to those mining company to produce metals to back that paper.

I dare not mention the names of these fantastic giants without their consent but we all know who they are! I would say the next time you meet an individual working for them to simply say Thank You for keeping this country alive.

'Mining Scams'

Age old professions for some of the less scrupulous miners are the scams to obtain investors and provide a great steady income for the pro-Con Man!

THE SMALL MINERS HELPER
A Basic overview of Mining and Processing in the US

These are generally prompted by a fantastic few assays, hundreds if not thousands of ounces per ton of the royal metals which would be in the head ore. Along with this are older (Not Current) Geological reports, usually performed by another mining company (which decided that it wasn't economically feasible) or other geologists who never completed the report and where no proof of cores and drilling could be produced.

To be a legitimate 'Good Deal' all the reports and records must be current or at the very least show connection by a timeline and be verifiable, assays sometimes thousands, have to be capable of duplicating the original results. Usually this cannot be done if there are sporadic breaks in time where ongoing work was not done.

Large Mining operations spend thousands if not millions of dollars and amounts of time from a few up to 20 years just to first prove an economic reserve before making a commitment to proceed with the rest of the phases required to mine it.

A sequence of operation starts with prospecting and a huge amount of money (for most of us) to prove the value of a reserve as you see in the sequence here.

1. Geological Exploration (prospecting on the ground or by aerial, and in most cases past history of some activity).
2. Government involvement in permitting to prove the reserves in the ground (Another expensive procedure)
3. Geological proof involving a drilling or sampling program, large paper trail, including records of depths, assay reports and a general synopsis of the value of the located reserve. Sometimes and in most cases engineering, permitting and building a pilot plant to prove feasibility of a process design, bulk sampling, and hundreds or thousands of assays
4. Plan of Operation, Mine planning engineering, Haul Roads, Milling and process engineer program.
5. Economic evaluation (to see if a profit can be generated by the final product).

The above can take from 1 to 20 years to complete and as you can see this money is spent before the second sequence is even started. This is the time where a company or individual decides to cut losses or proceed.

Again, know what you are investing in, milestones and time allotted before payback or losing an investment.

THE SMALL MINERS HELPER
A Basic overview of Mining and Processing in the US

Chapter 1

'Placer Mining'

A Sluice Box for separating the lighter gravel from the heavy gold these modern concentrators now engineered for optimal performance can sometimes using a few feet of sluice replace the long toms of yesterday.

A small 3 man placer operation

THE SMALL MINERS HELPER
A Basic overview of Mining and Processing in the US

Note the amount of material passed thru the trommel (L) this produced 8 Troy ounces of gold

The tailings were backfilled; topsoil replaced and is now grazing land for the landowner.

Placer Mining is basically the separation of values from Alluvial or Eluvial gravels, this can be done by many different methods using many different machines to capture the values as shown in chapter 2.

Placer mining is by far the easiest means of mining gold by small individuals and recreational miners. It can require nothing more than a shovel and a gold pan to large yellow iron equipment trammels sluices jig and pumps, all depending on the miner and their expectations.

Placer mining was very popular during the gold rush years of California. It gave an opportunity for the 'common man' to become wealthy yet again it made some poor men rich and some rich men poor. The most successful of the miners went on to be land and cattle barons and some just packed up and went back east. Mining before tourism produced a strong backbone on which this state was built. Anyway it is looked at, it made California a wealthy state even though this is hard to say since firsthand knowledge has proved California as one of the most unfriendly states in the seven western to mine for gold.

Many folks out there have the impression that the 'Ol Timers' got all the gold. Those of us in the industry also know they walked over millions to only pick up a penny. IT'S OUT THERE!

Another story is that the areas and claims have been over worked, no not really the claims which have been worked real hard might have had 1 or 2 men working a dredge over a 10 year period but remember they only worked different spots and may have only worked it for a few months a year. Q How about the bench gravel? A Never touched it with a dredge. How about the area over there where the real stream bed used to lie before the old timers diverted the water to work the claim? HMMM, 20 Acres is a lot of territory and as I said It's out there, just slow down and explore!

Understanding what to look for in a placer plays a large part of success if you're limited as to time money and health explore first. If you have all the equipment and don't have the limited time, explore first then work the richer areas to support the operation finishing with the lower grade afterward. Remember it's also good to do your reclamation as you go, the money runs out fast when the ground quits paying.

I have also been approached with, the old timers worked these claims all out, they have been tested to produce virtually nothing. MY Best Kind! 2005 I was told this very thing after spending a couple of hundred dollars doing some basic exploration, brought in some rented equipment, removed the overburden down to the bedrock,(picking up some small flakes close to) after opening got a portable electric jackhammer broke about 6" of bedrock to find many cemented cracks and fissures that produced a little over 7 ounces that day, average was a pound a week afterward.

Purchased the rented equipment and paid off the balance 60 days later. Placer Mining is smart work, not hard work.

THE SMALL MINERS HELPER

A Basic overview of Mining and Processing in the US

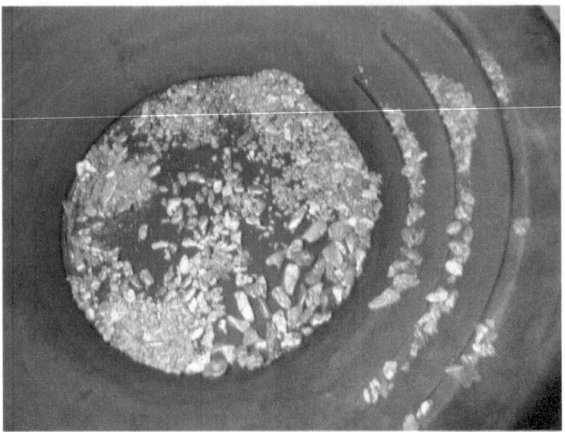

1st full week 11.5 Troy Ounces of cleaned gold from the bedrock

Chapter 1

'Hardrock Mining'

"They don't call it Hardrock cause it's easy"

The guts and glory of mining.

Hardrock mining is without doubt the most productive, yet most costly form of mining. Most small miners would attempt this by working an old mine or tunnel which has not been closed and has probably been abandoned. These mines are normally very old and extremely dangerous, as shoring if any may be rotted, mother nature and time along with geological anomalies may be occurring or have occurred over time. Extreme care must be taken to avoid rock bursts (caused by pressure)and loose or falling debris from its back (Roof of the mine) such things as excessive water inadequate or bad air also play a part in working these.

Outside ventilation or an exchange of air should be accomplished before even entering, this can be done with some PVC pipe and forced air into the mine, just remember when doing this that all combustion engines must be well away from the mine as it will draw the low lying heavy carbon monoxide inside to create a deadly atmosphere.

Thereafter shoring should be of utmost concern, before sampling or picking at any rock always maintain someone outside in case of mishap to give you a chance at survival should it occur.

Most of these mines still found were generally worked until the pay ran out or the old miners found a better occupation to support themselves. That doesn't mean that they won't still support a small production.

The large mining companies generally know where these occurrences of ore are, but either have already evaluated the reserve and or other statistics and have decided not to attempt any sort of production involving these mines.

Visible gold in some cases can be observed in some of these as in the picture above masqueraded by the large amounts of iron oxide, gold being very visible in the small quartz stringers intruding in this auriferous deposit located in Northern California.

The above particular vein structure averages assays of 1.3 Troy ounces gold per ton of rock. This is course free milling gold so recovery can be expected of 1 Troy ounce with a loss of .3 with the iron and microscopic gold not gathered on a shaker table. So to give you an idea of a ton of rock use this analogy 1 ton equals roughly 1 cubic yard or a 3 foot cube of the vein material has to be removed crushed and processed to give you 1 ounce of gold.

This does not account for having to remove the surrounding material to get the vein material which in some cases can be as much as 8 times that amount. Constant shoring and evaluation thru this process must also accompany the digging.

Hardrock involves a lot of extras such as hauling crushing and processing, but for the miner who wants to go year round, this is the ticket. With some simple assays you will already know what you have before you mine it.

THE SMALL MINERS HELPER
A Basic overview of Mining and Processing in the US

Chapter 1

'Prospecting for Gold and Silver'

One of the most important steps and the first is the prospecting, basically this is the ability of an individual to locate an area or structure which can produce a profit if mined.

Locating can be accomplished in several different ways or by combination thereof;

Locating areas by aerial and topography; looking at Topo maps, searching the web for images and maps showing outcrops drainages and likely areas which may produce a profit. Also looking at these maps for accessibility.

Ground search; Looking in known producing areas for indicators such as listed above

Historical Geology Research; Most areas of the US have been mapped and reports written by the USGS along with reports by institutions and mining companies. In this age most of the findings are published and can be found or procured on the internet.

Speaking with long time locals and other miners; Lots of tales out there but when listening most all carry weight.

Using the four tools provided above, a diligent prospector can gather enough information to start sampling of an area.

For Hardrock, Look for areas of high mineralization, outcrops of quartz with rust, and dark streaks of minerals. Veins inside rock, mines, and tunnels which seem to carry sulfides, different colors and mineralization.

For Placers, Known gold bearing streams, creeks and gulches. Look for areas carrying a high amount of black sands or other heavy minerals. Keep in mind not all gold is in the rivers, Check the bench gravels!

In the dirt, Quartz and stringer quartz carrying mineralization sometimes leading deep into the ground to pockets.

Hardrock; Once minerals are located in Hardrock one must sample from different areas to get a general idea of the production of the area chosen to be mined. This is achieved by sluffing off samples with a rock pick, evaluating them microscopically and/or by crushing the lot and chemically assaying and/or by fire assay. Just because you can't see doesn't mean it's not there! Many rich Gold and Silver mines do not show the metal as natural in large quantities.

Sulfides and Tellurides can carry as much as 30% of their weight in Gold or Silver. An example of sampling is shown below;

VEIN SAMPLING

VEIN OR VEIN STRUCTURE

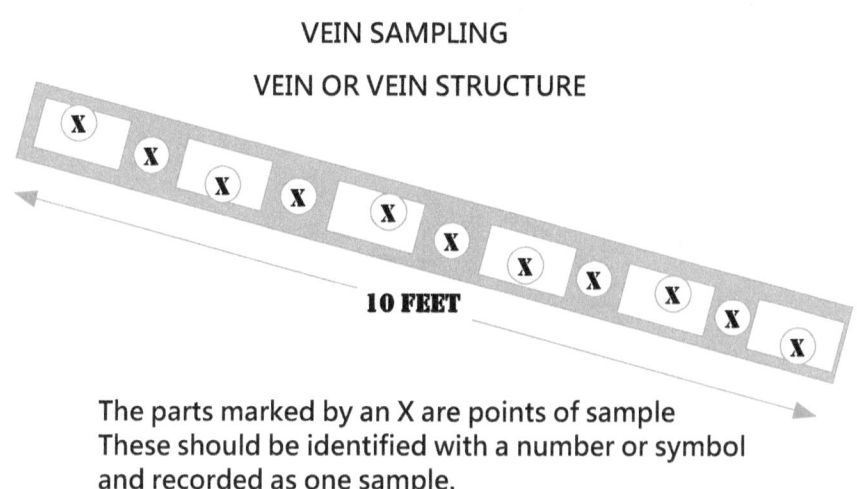

The parts marked by an X are points of sample
These should be identified with a number or symbol
and recorded as one sample.

ENTIRE SAMPLE SHOULD BE CRUSHED AND ANALYZED

Placers;

Areas to look for. 1. Inside bends of streams and rivers these seem to be the most productive for the person dredging. 2. under large boulders. 3. Areas which produce a lot of Black sand or heavy minerals when panned. Older bench gravels carry rich deposits as well.

When looking for a good placer one must read the land, cobbled or rounded rock opposed to jagged means the rock has been smoothed and tumbled by water. A lot of new miners think that the stream is where the gold is, this is not always the case.

If due diligence has been done in the locating part of prospecting you may find your original stream 5, 10, 15, or 30 feet from where it is now. A quick check of the area will reveal it's secret. The original pioneers often moved the flow of a creek or stream to accommodate cattle, mining, or for many other reasons, water diversion was not at all uncommon in past days as they did not have very much regulation at that time.

Water was diverted in some cases miles to a particular plot of land where someone built their homes raised their cattle or mined. The old stream or creek may have just been abandoned with all its riches intact. Keep this in mind when prospecting for placer, there's a lot more old gold than there is new.

THE SMALL MINERS HELPER
A Basic overview of Mining and Processing in the US

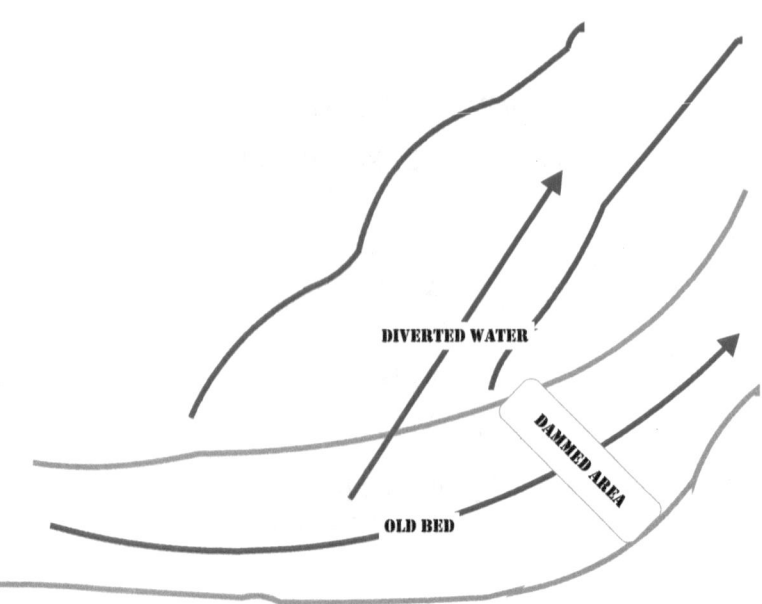

TYPICAL WATER DIVERSION BY PIONEERS

These old streams carry a lot of riches especially since most were not worked for the gold.

Chapter 1

'The Staking and Filing of a Mining Claim on Public Land as copied from the

California BLM website 2012'

WHO MAY LOCATE A MINING CLAIM?

Any citizen of the United States, a minor who has reached the age of discretion, a corporation, and non-citizens (aliens) who have declared their intention to become a citizen. (43 CFR 3832.1)

WHERE CAN I LOCATE A MINING CLAIM?

A mining claim can be located on federal lands (BLM and Forest Service) that are open to mineral location (entry). If you have a specific location site in mind, you may verify that the lands are open to mineral entry. This can be accomplished by checking with the BLM State Office Information Access Center (Public Room), who will assist you in determining if the lands are open to mineral entry by checking master title plats, records, files and other pertinent information.

WHAT TYPES OF CLAIMS/SITES ARE THERE?

Lode - A classic vein, ledge, or other rock in place between definite walls. A lode claim is located by metes and bounds. The maximum length is 1,500 feet by 600 feet. (43 CFR 3841)

Placer - All deposits, other than lodes. These include placer deposits of sand and gravel containing free gold and other minerals. Placer claims are located by legal subdivision. An individual may locate up to 20 acres with a maximum of 160 contiguous acres with 8 or more people (an association). A corporation is consider a single locator. (43 CFR 3842)

Tunnel Site - A tunnel site is where a tunnel is run to develop a vein or lode. It may also be used for the discovery of unknown veins or lodes. To stake a tunnel site, two stakes are placed up to 3,000 feet apart on the line of the proposed tunnel. Recordation is the same as a lode claim. A Tunnel Site can be regarded more as a right-a-way, than a mining claim. (43 CFR 3843)

Mill Site - Public lands which are non-mineral in character. Mill Sites may be located in connection with a placer or lode claim for mining and milling purposes or as an independent/custom mill site that is independent of a mining claim. Mill Sites are located by metes and bounds or legal subdivision and are up to 5 acres in size. (43 CFR 3844)

WHAT IS A LOCATION NOTICE AND WHERE MAY I GET THE FORM?

A location notice is a form that must be filed with the BLM, Local or State Office and your local County Recording Office. The following information must be included on the form; date of location of the claim/site, description of discovery monument, name of claim/site, legal description (metes and bounds or legal subdivision), and the names and addresses of all locators. There are separate notices for placers, lodes, and mill sites. You can obtain location notice forms from the BLM Public Room and stationary stores. (43 CFR 3833.1-2)

HOW DO I RECORD A MINING CLAIM?

You must file your mining claim/site location notice with the BLM, Local or State Office, within 90 days from the date of location of the claim or site and you must also file with the County Recording Office.

All **new locations** must be accompanied by the required fees of a $15 service charge, $34 location fee, and a $140 maintenance payment fee for the first year of location, for a total of **$189 per claim or site, after September 1, 2009**. Since we are after 2009, the fee is $189 per claim until further notified. Claims will expire September 1st, if you do not do your annual filings. Next paragraph describes your annual filing.

By September 1st of every year or before, $140 maintenance payment fee must be paid to the Sacramento office, or a Waiver with an assessment form and a $10 per claim assessment fee must be turned into the Sacramento office. (43 CFR 3833.1-2 and 43 CFR 3833.1-4) Sacramento's address is 2800 Cottage Way, Suite W-1623, Sacramento, CA 95825.

THE SMALL MINERS HELPER
A Basic overview of Mining and Processing in the US

MAY I CHANGE A LOCATION NOTICE AFTER I HAVE FILED WITH THE BLM?

You may file an amended location notice in order to change the name of the claim/site, clarify the legal description, or provide information that was incomplete. Also, you must file these amendments first with the county that the claim is in. After you receive the county stamp on the amended location notice, then file your amendment with the Bureau of Land Management State Office at 2800 Cottage Way, Suite W-1623, Sacramento, CA 95825. You may not add or delete locators or change the location date of the claim/site. There is a $10 filing fee for filing an amended location notice. (43 CFR 3833.0-5[p])

WHAT MUST I DO TO MAINTAIN A CLAIM?

Once a claim/site is serialized, an annual filing must be made on or before September 1, of each year to maintain the claim/site. If you have more than 10 claims, you must pay the $140 maintenance fee. If you have 10 or fewer claims/sites, you may choose to file either the maintenance fee payment or file the Maintenance Fee Waiver certification (a.k.a. small miners waiver). If you choose to file a small miners waiver, then you must also perform $100 worth of labor or improvements on all placers or lode claims during the assessment year (September 1, noon through September 1, noon). An Assessment Work Notice (Proof of Labor) form must be filed on or before December 30, along with the $10 filing fee per claim. For mill/tunnel sites, a Notice of Intent to Hold must be filed on or before December 30, along with the $10 filing fee per site. To learn more about mining claims/sites filing instructions, please visit our web page mining facts. (43 CFR 3833.1-5 and 43 CFR 3833.1-6)

WHAT IS A SMALL MINERS WAIVER?

A small miners waiver is short for maintenance fee payment waiver certification. A small miners waiver may be filed by those claimants holding 10 or fewer claims/sites, instead of paying the $140 maintenance fee by September 1, of each year. If you choose to file a small miners wavier you must also perform assessment work and file an assessment work notice by December 30, of each year. (43 CFR 3833.1-6)

WHAT QUALIFIES AS ASSESSMENT WORK?

Some of the activities that qualify for assessment work are construction and maintenance of access roads, development drilling and sampling, and buildings that benefit the claim. For more information about what qualifies as assessment work please contact your local BLM office.

HOW DO I TRANSFER A MINING CLAIM?

A mining claim is transfered by recording a Quit Claim Deed with the County Recorder where the mining claim is located, and then by filing the Quit Claim Deed with the Bureau of Land Management (BLM) State Office. The cost to file the Quit Claim Deed with the BLM is $10.00 per claimant, per claim. (Please call the County Recorder's Office for their fees). Quit Claim Deeds are usually found at office supply stores.

LAWS AND REGULATIONS GOVERNING MINING CLAIMS/SITES

Mining Law of 1872

THE SMALL MINERS HELPER
A Basic overview of Mining and Processing in the US

Federal Land Policy Management Act of 1976 (FLPMA) - See Section 314

Code of Federal Regulations, 43 CFR 3800

Interior Board of Land Appeals Decisions

WHAT MUST BE PAID OR FILED?

Pay the 2012 "Maintenance Fee" of $140 per claim, mill site, and/or tunnel site, on or before September 1, 2011, call (916) 978-4400, to obtain the **Maintenance Fee Form,**

OR

File a 2012 Maintenance Fee Payment Waiver Certification, Form 3830-2, (commonly referred to as the Small Miner's Waiver) on or before September 1, 2011. There is no charge for filing a Small Miner's Waiver.

Printable version of the waiver certification: Waiver Maintenance Fee Waiver (Note: Click on this and then you will have to click on the radio button left side to the word "Title," then in the box to the right type in the word "Maintenance Fee Waiver" then hit Search.)

AND

If you file the 2012 Small Miner's Waiver, you must ALSO file the following on or before December 30, 2011:

- For all PLACER or LODE CLAIMS listed on the Waiver, you must file a 2011 Affidavit of Assessment Work (Proof of Labor Form): Assessment (a.k.a. Proof of Labor), call (916) 978-4400 to order the **Assessment Form.**
- For all MILL SITES or TUNNEL SITES listed on the Waiver, you must file a 2011 Notice of Intent to Hold.
- A $10.00 per claim/site service charge is required when filing the Affidavit of Assessment Work and/or the Notice of Intent to Hold. The Affidavit of Assessment Work and/or the Notice of Intent to Hold must be filed on or before December 30, 2011.

Reminder: For those who filed a 2012 Small Miner's Waiver by September 1, 2011, you must file your 2011 Affidavit of Assessment Work on or before December 30, 2011.

CAN YOU QUALIFY FOR THE 2012 SMALL MINER'S WAIVER?

You can only file a Small Miner's Waiver if you own ten or fewer claims and/or sites nationwide and have performed $100 worth of assessment work on your claims by September 1, 2011. All claims, mill sites, and tunnel sites must be listed on the Waiver. The Small Miners Waiver must contain the original signatures **(preferably in Blue Ink)** of all claimants having an interest in the claims/sites. If an agent signs for the claimants, a notarized designation of agent, signed by all claimants must be submitted with the waiver.

WHERE DO YOU FILE?

Your local state office of the BLM

THE SMALL MINERS HELPER
A Basic overview of Mining and Processing in the US

Bureau of Land Management
California State Office
2800 Cottage Way, Suite W-1623
Sacramento, CA 95825-1886

Phone: (916) 978-4400 Information Access Center

NOTE: If you would like a date stamped copy of your document, you MUST provide an additional copy that can be returned to you.

BE ALERT! FILING REQUIREMENTS ARE SUBJECT TO CHANGE

As an owner of mining claims, YOU ARE RESPONSIBLE for keeping yourself informed of the changes in the filing requirements and the mining laws. Congress may pass legislation affecting filing requirements and, consequently, the procedure may change. It is suggested that you contact our office periodically to stay up to date. You may write us at the address above or call our customer service representatives at (916) 978-4400. You may also obtain from the Information Access Center guidance on locating a mining claim, additional forms, pamphlets, etc.

The use of the forms referred to in these Filing Instructions are optional; however, it is strongly recommended that these forms be used. Use of these forms will ensure that you provide the listing of all serial numbers, claim and/or site names, the original signatures required on the filings, and all other information that is necessary for you to complete your filing. These Filing Instructions and forms have been provided to you as a courtesy. To access the forms see the hypertext.

COUNTY RECORDER FILING REQUIREMENTS

In addition to filing with the BLM, file a 2011 Affidavit of Assessment Work (Proof of Labor form) or Notice of Intent to Hold with the county recorder's office. The location of this office will always be in the County Seat of the county in which your claims are situated.

NOTE: If you paid the MAINTENANCE FEE rather than filing a SMALL MINER'S WAIVER, indicate on your county filing that you have paid the 2012 Maintenance Fee to the BLM and the date on which you paid.

THESE FILING INSTRUCTIONS ARE PROVIDED TO YOU AS A COURTESY OF THE BUREAU OF LAND MANAGEMENT, CALIFORNIA STATE OFFICE.

Note; Each state has a BLM site and office for you to get all the correct information and paperwork to file, they do vary from state to state.

Below are examples to describe your claim.

Example BLM Lode mining claim

Example BLM Placer Mining Claim

THE SMALL MINERS HELPER
A Basic overview of Mining and Processing in the US

This information was provided by the website for the California Bureau of Land Management. **(BLM), in Sacramento CA.**

Chapter 1

'Mining on Private Property'

I know this all sounds good but, do you own the mineral rights? Is the area zoned for Mining? Do you own the water rights? Are you able to Prospect for Gold on your own property?..... Exactly!

This all seems like a good fairytale come true, but……….. The only one who can answer your questions is the county and state of which you live.

Some property purchases do not include the mineral or water rights they could be owned by other entities.

These are all questions that should be asked before making a decision to mine on your own land. There are many more restrictions than there are on public land.

Chapter 1

'Regulatory Agencies'

There are many regulatory agencies in the US which set the paces of mining. For the small miner working on public land the 'Set the pace' top agency is Bureau of Land Management, here is where all claims are initially filed and are the primary when it comes to regulations.

Next is the US Forest Service, many claims are located within their jurisdictional boundaries however all claim management is still controlled by the BLM.

Some states such as Idaho also use another agency called department of lands, and the DEQ (Department of Environmental Quality), to manage water discharge and air quality. An Idaho equivalent of the EPA.

All mining is done under the federal jurisdiction of MSHA, US Dept of Labor, Mine Safety and Health Administration. This entity seems to be originally set up to assure the safety of the Coal Miner where exploitation and horrible safety practices prompted the need for this agency. They also control Gravel Pits both metal and non-metal mining, rules are vague and fines are horribly stiff. In recent years small miners have suffered great losses due to non-compliance with this entity. However if a miner knows their rules and regulations better than they do, and makes every honest attempt to exceed them, one would not have to worry.

It is the miners responsibility to know which and how many agencies may affect your production, so it is a must to at least check out which agencies affect you and what permitting processes you may have to complete and regulation you are subject to before attempting to mine. Failure to do so could lead to losses of equipment and stiff fines, in some cases jail time.

Chapter 1

'Respecting the Environment'

Our forefathers weren't so friendly to the environment as we are today. Their quests for riches superseded their intelligence when it came to reaping the rewards of mining.

Use of large amounts of Mercury was quite common to capture and amalgamate the metals to be processed. Careless spillage and process control lead to the many associated deaths with this toxic heavy metal.

Mercury was heated to evaporation, released into the air which later fell to the ground to contaminate vegetation and water.

On just another note, the miners were and still are blamed for this huge contamination, but it must be understood that this carelessness was only a small portion of the total sources of mercury contamination as it exists even today. The only difference is that today mercury is now highly regulated in the US and many other countries.

There was and still is contamination by other natural sources such as forest, vegetation fires, and natural cinnabar (a mercury sulfide).

Manmade contamination is largely produced from industrial wastes and the burning of Coal which is used in the large electrical generation facilities. However these are not the only sources. Mercury is still carelessly used by some miners as a primary tool of gold recovery, but mostly in remote areas of countries other than the US where regulation and education of the hazards are mostly non-existent.

To give just a little better understanding why mercury is such a high concern amongst regulators and environmentalist, and which never seems to be fully explained to the public leaving a huge lack of understanding of this metal.

Liquid Mercury is naturally converted to methyl mercury in the aquatic system, rivers streams, lakes, and the ocean. Each step in the aquatic food chain that ingests the methyl mercury cannot rid itself of it. So when the small fish which has eaten it in silts or invertebrates which contain contaminated microorganisms, is eaten by the larger fish, consumed by the animal or human, it has each time been many times concentrated.

Large areas were recklessly mined with no reclamation, leaving massive areas without vegetation, and representing the face of the moon as in the historic site of the Malakoff diggings in Northern California where literally mountains of gravel were broken down by high pressure water cannons better known as hydraulic-n'

Today even the small scale or recreational miner is by public opinion and law required to put the area back as close to natural as possible. We as miners are still not trusted by the public based on the actions of our forefathers coupled with the lack of education and common sense of the average environmentalist and governmental enforcement agencies.

It is the utmost responsibility when mining on any scale, to create as little impact as possible on the environment, remember we live in our houses and we are visiting the homes of Bambi, Donald and Smokey the Bear. We are the true environmentalists and must demonstrate that we not only use the land for its resources, but are responsible stewards.

Rules to follow;

1. Plan before digging
2. Stockpile what you remove in separate areas to be replaced when you are finished, i.e. Topsoil with Topsoil Subsoil with Subsoil (This is termed as overburden)
3. Replace what you removed (the overburden) when finished in the order removed
4. Only re-vegetate with native vegetation, nature knows what grows best. (You can buy the recommended seed from your seed store, BLM can assist you in telling you what to re-seed with)
5. Control dirty water discharges from sluices. Use settling areas to filter out the sediments returning only clean water to the streams if used. Remember Bambi and the Fish would like a clean drink, as well as your neighbor 50 miles away.
6. If you plan to mine for a longer period of time put a small reclamation plan in writing (required by the USFS) keep it and follow it, keep your excavation areas clean free from debris and marked so that innocent people and animals may recognize any hazards which may exist in the area.
7. Treat the area where you work as your own Front yard!

"A good miner doesn't follow the rules because big brother makes them, but follows simple rules because they want to leave resources for tomorrow and pays back the land for the riches that was offered them".

THE SMALL MINERS HELPER
A Basic overview of Mining and Processing in the US

Chapter 2

'Placer Mining Equipment'

Safety in Placer Mining

Not enough emphasis can be placed on Safety. There are many hazards in placering normally not discussed by the small miner, but here it is!

Most of the individuals and small miners when serious will look and mine in areas remote to our conventional normal surroundings. Issues such as no phone or cell signals are common.

Rugged and steep terrain, loose or jagged rocks, unpredictable weather and the inability to self help in case of an emergency, all play a part in pre-planning to keep safe and alive.

Perfect examples are a man or lady suffering a heart attack or heat stroke after the vigorous activity exerted mining. Without a plan and time having a huge factor on the outcome of the person's health will almost certainly lead to disaster. Taking for granted that a fall won't happen, or the weather won't change is almost a certain recipe for an incident.

The raw truth is how do we react when we can't reach 911, when the road suddenly becomes so slick that our vehicle can't make it out, we fall striking our head or breaking a bone without the presence of a partner to help, a mountain lion decides he needs lunch....Then what?

Prior planning plays the only part that can stop an incident into becoming a catastrophe.

Here are some tips on how to be prepared for the emergency when it happens.

1. May sound stupid but, address all worst case scenarios in writing, keep them with you and follow them, they will help prevent confusion which may lead to panic and mistakes, Ol Murphy, he's out there!

2. Find a way of outside communication, radios may work well with agencies that can help you provided you can reach them on their frequencies. If no other way to get outside communication, get a satellite phone. Never be alone!

3. Always let someone know where you intend to be i.e. GPS coordinates work great, if not give them a map of how to get there and let them know what your deadline is for you to return, so that they may call for assistance if you do not return by the that time.

NEVER MINE ALONE!

Chapter 2

'Placer Mining Equipment'

The Sluice Box

One of the best instruments of concentration in Placer mining is the sluice box. Hydrodynamics, Mechanical engineering all in one simple looking package.

Shown in this picture are a row of Sluice boxes in line to capture even the smallest particles of gold.

Sluice boxes can be as simples as a wood or concrete trough with carpet laid in to the most sophisticated designs as we see today in some of the manufacturers of these. Using Water dynamics and mechanical engineering to immediately suspend the lighter materials and capturing the heavy, (such as gold).
There are many manufacturers and designs out there but one of the most popualar are the ones using the Hungarian riffle, whereby thru hydrodynamics a vortex is created as the water and lighter materials moves over the riffle capturing the extremely heavy material and holding it to the bottom of the box.

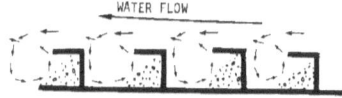

Sluice boxes must be cleaned periodically to prevent the loss of heavies.

Normal drop for a sluice box is about 1" per foot in length; this is how they are designed to work their best clearing out the light material by friction of the water very quickly.

All boxes should be cleaned if using a single when the heavy material is seen being stacked in the riffles beyond the 3rd riffle to avoid loss.

Feed into the Sluice should not exceed the ability of the water to remove the lighter material within 45 seconds to a minute. If sluice does not do this angle may be too flat or water flow may be inadequate, adjust as necessary or follow manufacturer's instructions.

The most loss of gold from a sluice is from 4 factors (1.) Ignoring (2.) Tampering (3.) Laziness and the (4th) greed.

Addressing number one; constantly running material thru the sluice and ignoring it's operation can lead to significant losses of high grade material. You need to see if large rocks are stuck in the box and clear them, that material is not 'Packing' the riffles, where they no can perform, and that water flow is sufficient, without the attention to all of these factors its almost positive that you will have significant loss.

Addressing number two; Playing in the sand box is prohibitive every time you stick your finger in, the hydrodynamics and engineering is disrupted, vortexes collapse releasing high grade material leading to the loss if not captured further down the box. Other than clearing an occasional large rock (As this also does the same thing as sticking a finger in) the box should be left to do its job.

Number three; Well it is what it is, riffles fill gold has nowhere to go but out!

Number four is very important; If you see the riffles starting to load past the third, 'It's Time' you will get less by packing than you would if you took the time for a clean out the loss of time will not equal the loss of material.

Chapter 2

'Placer Mining Equipment'

The Highbanker

The good old high banker, basically just a sluice with a pump to allow the constant addition of material and using an outside source of water to wash the gravels and replenish the flow of water. Small units are used in recreation generally with a hopper and wash area dropping into the sluice for concentration.

These are of great use in areas without large amounts of water as a recirculation setup can enable constant processing.

The high banker can be used as an onshore method by attaching a dredge hose to it pumping the material thru the sluice for more control of the concentrate.

This piece of equipment can also be used in conjunction with a trommel, screener or a jig for a better cleanup activity.
equal the loss of material.

'Placer Mining Equipment'

'The Gold Dredge'

This Bucket Style Gold Dredge is located in LaGrange CA.

Constructed at the cost of $543,148.00 in 1933

Today common use is of the suction dredge whereby high pressure water is injected into a venturi causing a very high degree of suction to remove gravels in a bed of water, pumping them into a sluice or concentrator. Commonly used by recreational and small scale miners globally.

These suction dredges come in all sizes of inlet with the most widely used being the 4" dredge because of the ability for one man to move in and out of harder to reach areas.

There are now quite a few manufacturers of this equipment all have been engineered well and designed to perform. Features such as easy assembly and disassembly, hookah, fresh air pumps, and wet suit heaters are also a great accommodation offered by many of the manufacturers to compliment these machines.

A good choice for someone who prefers working in the water, opposed to working on dry land.

A 4" Portable Suction dredge

Seen here a professional 'Dredger' and friend, gives instruction to a peer not shown.

His 4" Suction Dredge is pictured in the background with a full hookah setup.

Chapter 2

'Placer Mining Equipment'

'Jigs and Trommels'

There are many type shapes and styles of Jigs they operate by shaking, cleaning and stratification of the material being fed into them, the lighter materials going over the discharge with heavy material moving thru the screen and collecting at the bottom.

THE SMALL MINERS HELPER
A Basic overview of Mining and Processing in the US

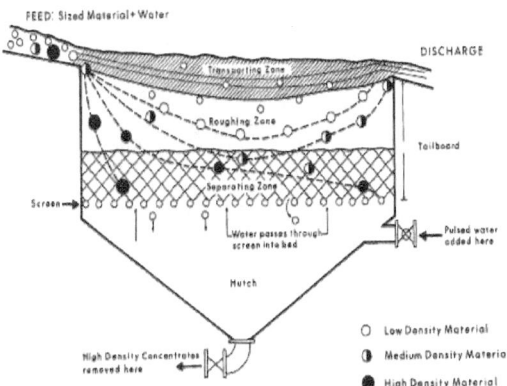

Trommels are also all shapes and sizes from some of the mini's offered by recreational shops up to and including the 5' X 40' shown below.

Pictured here in the rear is a larger full screen trommel 36" X 60" mounted in a trailer

A smaller 10" X 60" portable unit is pictured in the front.

Trommels are considered wash and screen plants, as the material enters the barrel it is constantly beat and agitated expelling the smaller cleaned material into a sluice or concentrator for better recovery by washing and breaking apart all of the clays and dirt which might otherwise carry the values off with the larger rocks and cobble. Screens can normally be changed to fit the user's needs and expected size of recovery.

Both Trommels and Jigs are considered wash plants.

'Placer Mining Equipment'

'Digging and Excavation'

Hand digging is one thing you will find yourself doing frequently as if even using equipment, there will be cases where hand work is required.

One thing is for sure, hand digging is the most accurate way of pinpointing a targeted area to obtain good material. Sometimes and most times difficult dealing with dirt rock and cobble such as a placer. However some of the work can be a little less cumbersome by using the right tools.

Shovels can be altered to enter the gravels by grinding them into points, I have even seen teeth welded to them similar to the teeth on a backhoe bucket. A little creativity up front saves a lot of sore backs.

Picks and bars can also be altered to fit the need, digging anywhere with a dull shovel is tough, taking some hints and lessons from excavation companies and ditch diggers which really do a lot of hand work and finding out that alteration and creativity really saves a lot of hard work. Using things such as a probe can also allow you to locate large covered boulders which you may want to dig around.

One thing to always take into consideration is gravity, should you dig out a large boulder as you dig under it may shift sometimes pinning the digger having things such as 2X4s, 4x4s and some plywood along with other wood products for shoring will reduce the unsafe nature of an excavation.

When digging into wet or moist ground the drying action (Depending on the dirt) may cause material to collapse into your hole, therefore shore it up and you will find it easier to access especially if having to go back later.

When digging with any kind of equipment Trenchers, Backhoes, excavators beware that this equipment is heavy and can collapse a trench or excavation if not properly shored. Another thing to remember is that if your ditch or hole id deeper than knee high start some shoring as if it is chest high or deeper collapsing of the sides can restrict your chest where you will no longer be able to breath, dirt and rock are very heavy.

Always plan out an escape route from an excavation, in smaller holes cutting steps in with a shovel is a good idea at each end of your dig, ladders may be necessary in deeper excavations. This also aids in getting in and out many times as most of us do.

For the small miner a quick rig of a tripod over the hole with a block and tackle or pulley device will save a lot of trips up down in and out where by a bucket can be pulled up out of a hole and deposited on the top. A buddy system works well when removing the paydirt by hand. Tools and creativity mean everything to the small miner.

'Placer Mining Equipment'

'The Grizz' or 'Grizzly'

The Grizzly is a formative way of discarding and separating large unwanted rock which wouldn't normally fit in your hoppers, Trommels or sluices. When digging by hand or equipment nothing is more cumbersome then having to unplug something because too large of a rock was dumped in. Grizzly's can be made in all shapes sizes and will just about fit anywhere you want one.

This equipment is commonly used by road crews and mines to separate their large rock from the useable dirt and gravel. You see them mounted on all types of equipment and free standing on the ground.

The grizzly

Note the Large grizzly at the Head Hopper of the Trommel

'Placer Mining Equipment'

'Crushers'

A Partially disassembled 1935 era Jaw Crusher

A view of the same jaw crusher from the top

THE SMALL MINERS HELPER
A Basic overview of Mining and Processing in the US

There are many crushers used by the small miner, the most popular being the Jaw and impact crushers. The jaw works by a twisting compression of the rock where by an impact works by slamming the rock into a solid face pulverizing it from the high speed impact.

This is an old commercial impact mill that still works today.

It has 4 hammers and can be run wet or dry.

These crushers are mostly used in Hardrock mining but on occasion known to be used in sluice gravel after concentration to release any of the finer material which may still be present in them. Below are some pictures of homemade equipment that works.

Jaw Crusher

This unit would normally be used as a lab crusher but for a small miner would crush 500 to 700 Lbs. an hour depending on hardness of material.

Roller Mill or Roll Crusher

These mills are primarily used to pulverize finely crushed rock turning it into whatever size is desired there in this unit there are 1 set of rolls the one roller is adjustable to size the output. Some have multiple sets of rolls to slowly decrease the size to a fine powder.

Large adjustable roll crushers are commonly used in the diamond mining industry to reduce the Kimberlite to a size of the most common size of diamond it produces.

Homemade Ball Mill

The Above equipment saved the miners who made them thousands of dollars but in some cases took large amounts of time to design weld and assemble them.

I personally witnessed most of this equipment in operation some years ago, and it did work just as well as their commercial counterparts.

If one has the time a little ingenuity and planning can same thousands of dollars in equipment costs.

THE SMALL MINERS HELPER
A Basic overview of Mining and Processing in the US

Chapter 3

Hard Rock Mining

Speaking of Hardrock mining, I don't know if they meant the rock was hard or the hard work involved in removing the values. If I were to be asked from experience, I would say both.

However, when using this term in the manner of indication, it generally refers to the removal of ore in its native state. This can be done by many means; today most large mines are open pit mines where vast tonnage of material is removed to access the valuable ores.

In general when we speak of it as small and artisanal miners, we refer to tunneling in on an outcrop, opening a past producing find, or simply working an old mine where values still exist.

Safety in Hardrock Mining

The very first thing I must say about Hard Rock mining is that it is dangerous. Much knowledge and research should be gathered before attempting to even enter an old mine, or doing any digging where one might place one's own body.

The number one rule of thumb is to NEVER walk into an old mine without the following.

- A presence of another person on the outside of the mine whose job is not to try to rescue you in case of mishap, but to go for help. (**You must always have someone present on the outside to go for help, this is absolutely a must!**).
- Have an air monitor at the very least an oxygen detector, a single gas Oxygen detector can be purchased as low as $120.00 this is a must as the air we normally breath is about 20.94% (O_2) oxygen and air below 16% (O_2) is considered dangerous, (Oxygen deprivation), I personally get tunnel vision below 17.5%. Better multiple gas detectors are available, and should be considered if you suspect any other gases in the mine and definitely if a vehicle has been running anywhere near the mine for any length of time.
- Have plenty of Light to last 1 hour more than your intended stay in the mine.
- Have plenty of rope (a harness is also a plus if available)
- Wear some sort of head protection.
- Have some trail flags to drop if you plan to go in far.

The lists above are only basics, it's simple, if you are missing any one of these items stay out!

A person intent on Hardrock mining must learn the terminology and basic technology to proceed with mining a value. In most cases a small or artisanal miner with limited funds and tools must learn to improvise and 'Hi-grade' the valuable ore without diluting it with waste rock. They must always be highly alert and attentive toward their surroundings.

Sounding the rock and looking for high stress areas, cracks or signs of weakness which might cause a spalling or drop of rock from the back, face, or walls (referred to as the ribs) of the mine, these areas need to be scaled and cleared of hazards before mining.

THE SMALL MINERS HELPER
A Basic overview of Mining and Processing in the US

Rock Bolting, shoring, and/or cribbing may be required to insure the area can be safely entered and mined. Barring down and scaling is a constant, "This is an ongoing effort the entire length of the miner's workings, especially in areas disturbed by any removal of ore and rock".

There is an excellent publications on procedures for this process on the internet http://www.dmp.wa.gov.au/documents/Guidelines/MSH_G_UGBarringDownAndScaling.pdf and thru educational institutions such as Edumine, UBC and Colorado school of mines.

Tools

Blasting without a permit and license is not an option, not only is it dangerous, but may be illegal in most cases (Most information can be obtained from the Bureau of Alcohol Tobacco and Firearms website, or you can call their local office. There are many expandants sold that can be poured to do the same job with instructions available on how to drill and use them (such as dexpan). These expandants will swell and fracture the areas you want to remove, the same as explosives and without many of the dangers and hazards associated.

Buckets are a small miners best friend, they allow quick cleanup of any debris you might want to remove from the mine or your working area. Some sort of rubber tired cart, or wagon is a plus for ease of removal without lifting heavy buckets, a wheelbarrow is not recommended as the floors of a mine may be uneven causing loss of control.

Good working lights and a professional miners light with a 12 hour capacity are a must, good rubber boots (since most mines seem to produce water), along with good leather gloves.

The pick, and shovels are a necessity, along with a good quality rock pick, single jack, a 5 to 6 ft. digging bar (the type with a curved flat and a pointed opposite end) and a scaling bar.

Drills and jacklegs are a definite plus, but must have air or in some cases electricity to run them. Remember if you need power equipment, and have to run gas or diesel to power them. DO NOT place them in close proximity to your mine as you may bring carbon monoxide (CO) into your mine. Many small and artisanal miners have died from this, even having what they thought was good air ventilation. If in doubt always have a good 4 gas detector to alarm you of any gasses which might enter the mine. Note also that MSHA requires a self-rescuer to be on your person when entering a mine, these can be purchased at a mining equipment supplier, they work by conversion of Carbon Monoxide to Oxygen allowing time to escape a high CO area.

A good ventilation system is definitely necessary in some pocket mines where no ventilation shafts or winzes were ever opened. This can be done by fans and bags or by PVC pipe with a blower attached, again remember the carbon monoxide factor.

Dust masks are without doubt beneficial, but remember to check them often and change as manufacturers recommend. Some dusts enter the lungs and can cause great health problems further down the road.

The lists of available tools for Hardrock mining are numerous, but using the basics along with some common sense will make you profitable. Do a little research since new and easier to use tools are always being sold, each mine each deposit and each ore always seems to be unique, so using some creativity always pays.

If at all possible having a small crusher at the mine always helps, this allows you to crush small samples or even your ore before transport.

Bulk Sampling

One of the most important parts of Hardrock mining is to take samples for assay. A good bulk sampling program is a must, allowing you to know what, where, and when to mine.

Taking a Bulk sample means that you are going to take multiple samples of roughly the same size from an area you plan to mine so you can determine roughly (and I do mean roughly, I will explain later in assaying), what to expect as far as values from your mining.

(AN EXAMPLE)

I have a 10' face I plan to mine there is a 2' vein exposed, Almost horizontal, I see a rusty composition on one side where the quartz stops and turns into ultramafic rock and a very dark composition on the other side where it turns into another rock I can't as of yet identify (These are known as contact zones); I see visible gold and some sulfides in a darkened quartz matrix in the middle.

I will need samples of this taken from 4" above the top to 4" below the bottom (This way it includes the contact zone and any waste rock that will dilute my processing after mined) roughly 6 inches apart top to bottom and 12" to 24" apart the entire length, each sample should be drilled cut or broken out to whatever depth you intend to mine. Each spot should be identified and numbered, along with the ore removed. Ore bags are definitely a good idea. I will retain one part of each sample, I will send one part for assay of each sample, and I will take one part of each sample crush split and send 1 lb. of this out for assay. (Since assaying all the samples may be too costly the split sample would give a representation of the whole ore you intend to remove).

Now this may seem like a lot of work, but this is the only way you can decide what to take out and where to take it out from. Most gold and silver is microscopic and the areas in which visible metals cannot be seen may actually be of more value than the visible gold area. The visibility is only a good indicator that the values are present.

The Assay

The assay, commonly referred to a the Fire Assay, is a centuries old time proven scientific method of collecting the Nobel metals in a lead compound, then by cupellation oxidizing the lead into a PbO and absorbing it back into a bone ash or magnesite cupel leaving a bead of noble metals which is weighed and parted using a Nitric acid compound, then reweighed and recorded for its percentages of the Nobel's.

As stated above a time proven method, 'that in order to sell it you must be able to cupel it'.

Accuracy of a fire assay, according to statistics, the fire assay when properly done is the most accurate means of figuring the weights and percentages of a sample. A controlled sample provided by the miner, geologist, or their representative is given to an assay laboratory where it is carefully weighed and recorded, prepared by pulverizing, split down to a 29.16 gram sample (1 assay ton) then fired with a Litharge flux and reducer causing a precipitation (raining effect) thru the ore capturing all the metals excluding iron compounds into the lead metal compound.

Afterward the crucible is removed and poured into a mold containing the slag glass and a lead prill weighing roughly 30 grams. The lead is then removed slagged off pounded into a cube and cupelled at roughly high enough temperature to ward off or oxidize any base metals. What is left is a Nobel metal bead containing only gold silver etc. This bead is then carefully weighed on a scale and recorded. (a bead weighing just 1 milligram equals 1 ounce troy of metal per ton). The bead is the parted in an acid compound to correctly determine percentages and then recorded for reporting.

The up side of fire assaying is that when properly performed under controlled conditions the assay is rarely wrong. The down side is that it is only representative of the sample received by the lab. If the assay sample has been compromised by any means of gathering, handling, or improper methods of take, it may not be representative of the ore body. It is extremely important that anyone looking to get a good assay understand their part before it is sent to a lab. One assay of one sample is only representative of that sample, and if too small the sample could contain a tiny sec of gold which will lead to what assayers call nugget effect and be totally off. Assays should always be multiple samples of the same ore to eliminate this, and a compound assay of 6 of the same sample will generally eliminate nugget effect, but do note this anomaly does occur frequently especially in free milling and placer ores.

Chapter 4: Milling and Processing

BASIC EQUIPMENT.

A Grizzly, Hopper, Primary Crusher (usually a Jaw, Cone OR vertical crusher) are used to take the Large rock down to manageable sizes to be further crushed to a 2" minus. Then it is normally taken or conveyed to a secondary crusher which can also be a jaw, cone or impact to reduce it down to a 3/8" to ¼" minus where it can then be sent to a SAG, Autogenous Mill and or a Ball mill depending on size of the feed, for grind and re-grind.

From this point it is taken or pumped to either a gravitational circuit or flotation circuit or both, depending on size and composite of mineral particles to be extracted, (most Gold and silver cannot be seen with the naked eye).

Autogenous Mills are normally mills which rotate larger rock from the crush circuit, mine or in any fashion used throw and tumble this rock by use of lifters or large bars mounted inside the mill to lift throw and tumble, using the rocks own weight and tumbling action to break it apart.

Semi-Autogenous mills or SAG mills are virtually the same large fashion of mill with balls, (usually a charge of 8% to 21%) added to supply a crushing grinding action. These mills (Both the AG and SAG), are generally large in diameter and short in length and are generally always used in conjunction with a ball mill for regrinding the ore into smaller particles.

Ball Mills come in all different shapes and sizes but are virtually all the same as they are a machine designed for fine grinding. Many small miners and artisanal operations use things such as barrels, old propane tanks, and custom fashioned pipe to achieve the rolling grinding action of a ball mill, some are powered by large motors some are even turned by hand with a crank. In some cases even hard cobble is used where balls are not available. Ball mills are usually characterized by its length usually being 1-1/2 to 2-1/2 times its diameter.

Ball mills normally operate with an approximate ball charge of 30%, normal rotation of this mill is 58 to 62 RPM, with most ball mills using a water ratio of 35% for grinding. The crushing and grinding of all of the above is commonly referred to as 'comminution of ore' in the mining industry. Ball mills as stated above have both a crushing and grinding action created by the liner within which have lifters to, at a certain RPM throw the balls to crush, and a rolling weight of balls to grind the ore. A very efficient machine.

Illustration Ball and Sag Mill

Ball Mills in general > Length Exceeds Height

SAG Mills in general > Height Exceeds Length

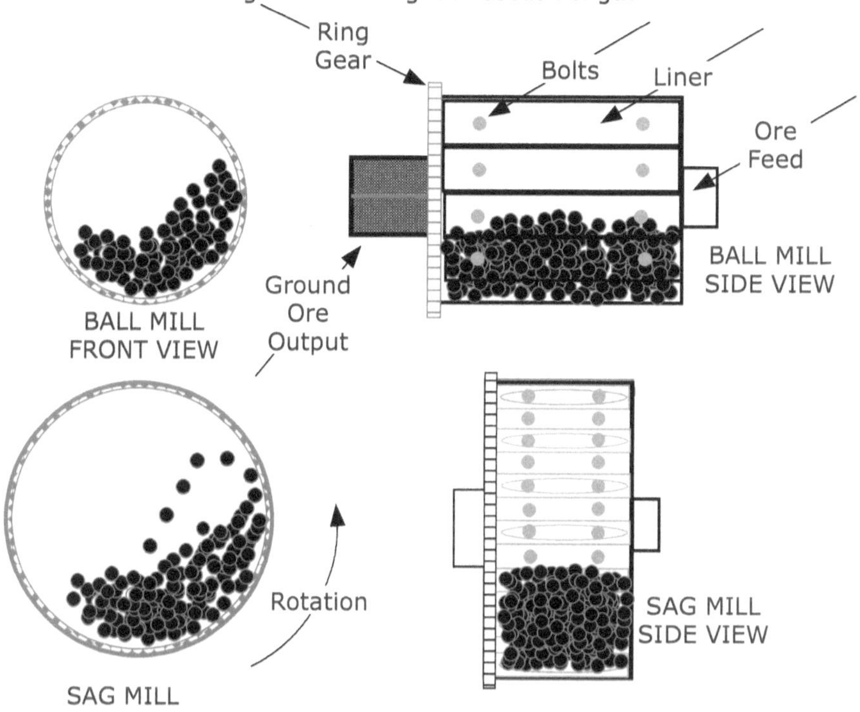
Ring
Gear
Bolts
Liner
Ore
Feed
BALL MILL
SIDE VIEW
BALL MILL
FRONT VIEW
Ground
Ore
Output
Rotation
SAG MILL
SIDE VIEW
SAG MILL

DANGER

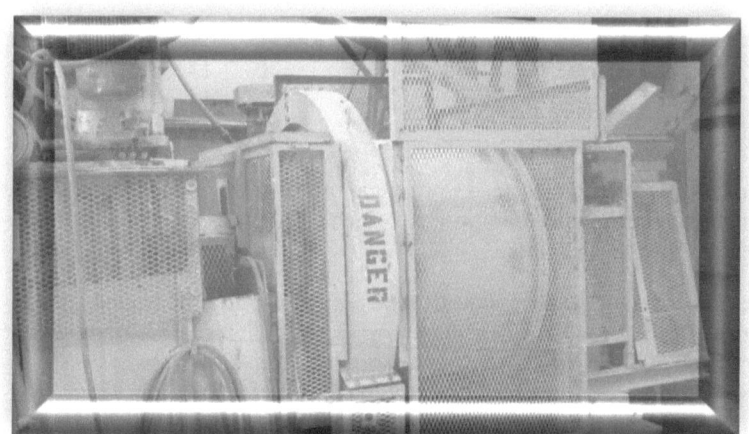

Ball Mill and SAG Mill Illustration

Smaller Marcy Ball Mill

The Ball mills used in most operations are sized from huge multiple units capable of grinding Hundreds of tons per hour down to small commercial units and The "Cocos" used in Columbia which do only several Lbs. per hour.

These units are generally sized to match the output tons per hour of the operation. Ball mills almost always work in tandem with a classifier screen. The material passing thru is sized to the optimal concentration capability of the equipment (down the circuit) that actually does the concentration, i.e. Centrifugal and table concentrators for gravitational and/or Flotation cells for using the flotation method of concentration.

When speaking of the classification screens, the material that passes thru the ball mill comes out at various sizes, if the screen is sized per example at 80 Mesh all material less than 80 mesh will pass thru the screen for concentration (generally referred to as the P-80, or the 80% smaller than that screen size). The remainder will be conveyed back to the ball mill for re-grind and will not be released for concentration until it can pass thru that screen.

Gravitational Systems

The next step if gravitational, (Generally used for free milling Gold, Placer, and Some heavy sulfides), is a visit to the primary concentrator, this can be a centrifugal such as Falcon, or Knelson concentrators (Even though many more are manufactured) these all have riffled bowls which turn at high gravity to separate the light materials from the heavies. They can be either a pass thru system or batch unit. Below is a Knelson and a cutaway look to reveal how it works.

Knelson Concentrator

THE SMALL MINERS HELPER
A Basic overview of Mining and Processing in the US

Cutaway view of a Knelson batch Concentrator (A personal favorite)

Generally speaking some processors may use primary shaker tables, such as Diester, Dove, and some such as US manufactured GMS in lieu of a Centrifugal system. Whether single or multiple, it is the preference, costs, and tons per hour, which mandates this.

There are many different ways to set up a complete circuit, a lot of study, time and effort should be put into this before purchasing any equipment.

Next is the finishing step, there are too many finishing tables on the market to say which ones are best. My personal preference is the RP-4 manufactured by Global Mining Systems and the M-10 by Action mining, both manufactured in the USA, as they are inexpensive easy to maintain and will produce well with minimum experience and a little bit of training. The Gemini (by APT processing) and Xtruder (manufactured by Madden Steel), finishing tables are excellent for very fine gold recovery at the finishing end of the circuit but a little pricy for the small operator.

Flotation

For the High Volume producer Flotation has been for historically, the preference. Although this method is not as environmentally friendly since the use of chemicals are involved.

It seems to be the long standing way to separate and be selective of each mineral a producer is trying to concentrate. Initially this method is very expensive requiring extensive testing to achieve good results. This method also being the only hope to concentrate some of the more complex ores as produced in some of the Western United States.

Flotation circuits generally use sets of cells to fit the tons per hour (TPH) one is required to produce. There are both mechanical cells and column cells, with the column being more the trend these days.

Generally a typical flotation system consists of a set of rougher cells, a set of cleaner cells and may even consist of scavengers to refloat what might have been lost in the initial concentration attempt. Some systems even configure gravitational separation into the same circuit.

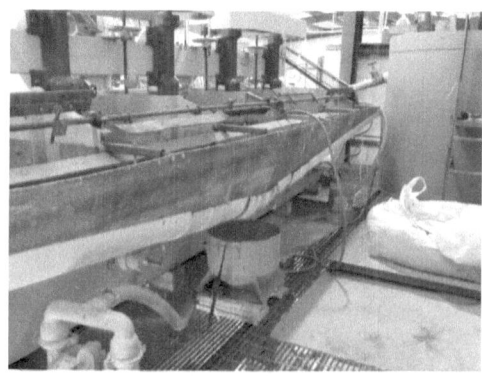

Typical Denver Rougher Cells (Mechanical system)

The objective of flotation is to make the wanted minerals float, to be scooped off with the paddles seen above. To do this one needs to make Hydrophilic particles (Particles, which like water) into Hydrophobic particles, (Particles which like air), so they may attach themselves to a bubble for a ride to the top of the water, at which point the froth will hold them up till scooped off. These are the concentrates.

This is done by chemical reaction. Collectors such as Xanthates are used with promoters such as X-523 by Prospec chemicals for selectivity of the mineral, Frothing agents such as Tennafroth 250 frother and Methyl Isobutyl Carbinol as a froth stabilizer, Pine oil and Diesel are also used to create a froth and stabilize it. Most mining chemical companies can supply more information on what to use and recommended dosages to be used.

However even the small operator can scale this down to a size to use this method. There are many books sold and online on Froth Flotation, One of the best I've found is... The Chemistry of Gold Extraction by John O Marsten and C. Lain House.

Leaching of Concentrates

There are so many methods of chemical leaching processes that they can only be touched on here in this small book. One thing for sure is that up until today one of the best proven methods is the use of Sodium Cyanide, It has proven itself over many years in heap leaching processes big and small worldwide, even more today in batch systems.

There are several problems when considering the use of this chemical, the first being social acceptability. The next thing that must be addressed is the governmental permitting, before cyanide can even be legally purchased. (For information on permitting call the EPA). Please note this is not something you do in the back yard. It takes a lot of experience and/or understanding to safely handle this chemical. Allowing this chemical in solution, to drop into an acidic pH will create HCN, or Hydrogen Cyanide gas (Very Lethal) not to mention losing the values and monies invested to start this process, endangering not only you, but the environment as well.

There are many backyard scientists that may say otherwise but this is not something to experiment with, if you are without the proper education or knowledge to control the reactions.

Today most small miners use an agitated batch system, closed loop, using large tanks with Water, Sodium Hydroxide, Concentrate and injected Oxygen to extract the values from the Concentrate, adding either carbon columns as a collector (By using the carbon columns it adds an extra step to the process as the carbon has to be stripped of its values in a hot caustic solution to obtain the values collected) or pumping the pregnant solution directly to an electrowinning system to be plated out as sludge on stainless steel plates. After which the plates are washed sludge collected and melted to a dore. It normally takes from 24 to 48 hours to complete a batch. After the solution is barren it is returned to a barren holding tank for re-use again. The water pressed from the washout of the plates is sent to a tank or holding vessel for destruction. (For more information about CN destruction, try checking out the **INCO** destruction process or the SO4 destruction of cyanide.

Destruction is one of the most important features in cyanide leaching as no water can be released into the environment until the WAD and Total cyanide values have been depleted to the values mandated by the World Bank, and EPA, generally 0.05 mg per ton WAD and 0.20 mg per ton, for total CN for release into the environment. For the best information regarding this subject check out The International Cyanide management code (ICMI) www.cyanidecode.org .

To the dismay of many who read this, noted are many different leaching chemicals which have been tried and proven in the laboratory over the years such as Chlorides, Halides, Acids, Thiosulfate, Thiorea, Bacterial, etc., the problems are numerous with these leaches, such as controllability, failure on a large scale and in most cases being cost prohibitive, and without a good knowledge of chemistry one is setting themselves up for failure. These methods although proven on small scale and laboratory materials have been tried by the large mining companies spending extreme amounts of revenue only to find them cost prohibitive and only to return to Cyanide, as a leaching agent. Not to say that these other methods may become economical at any time, but also remembering that they are chemicals, some either highly reactive or toxic and since all will contain metals, and all have some method of treatment associated with their use to be released back to the environment.

Chapter 5: Smelting and Melting

The terms Smelting and Melting are vastly different. One must note that the term Smelting refers to a pyrometallurgical function of chemistry using high temperature flux reagents, gasses and time to reduce an ore to a metal state.

Whereby the term Melting is basically melting a pure metal to a fluid mass to pour it into a mold or ingot.

Note that both will require a certain flux to either prevent oxidation in the molten stage but melting does not require much more than heating a metal to slightly above its melting temperature to be able to pour into a mold.

Smelting is a process used primarily because most concentrates will not exceed the 33% metal mass needed to allow all the metal to collect together to pour into an ingot.

Smelting however requires chemical reactions at high temperature to oxidize unwanted metals along with collector metals to achieve at least a 33% metal mass of the wanted metals, reduce and collect them in that mass then pour them into a mold to be refined.

Unless the mass of metal equals 33% or more all the wanted metals may not collect (drop) and may be left suspended in the glass (Slag).

Below is some examples of this collection procedure by using the smelting process.

Basic Smelting with Copper

Soda Ash

Anhydrous Borax

Calcium Fluoride

Activated carbon charcoal (Coke)

Virgin Copper in powder/grain form

Flux components:

The following may be adjusted accordingly:

- 16 oz/ 10lbs ore crushed to -20 Tyler mesh (or finer)
- 8 oz /5lbs soda ash
- 4.8 oz/3 lbs Anhydrous Borax
- 3.2 oz/ 2lbs -calcium fluoride
- 3.2 oz/ 2 lbs Activated Carbon charcoal
- 5.6 oz/ 2lbs , Virgin Copper in powder/grain form
- 2250F for 3 hrs

Melting Verification procedure—

- Proper capacity Crucible furnace with adjustable temperature control (Electric or Gas)
- Mixing apparatus for flux, ore and copper
- Carbon rod for stirring
- Mold

Samples will need to be crushed to 20 Tyler mesh

Below are instructions for melting 1lb of Sulfide containing concentrates with flux and 5.6 ounces of copper, (or 35% pure CU for collection), then pouring a CU/AU/AG dore bar.

1) Have samples crushed to -20 mesh or finer and prepared for processing

2) Load <u>only ore (Cons)</u> into furnace alone and **roast** ore (Cons) at 1100°F for 4 hours prior to firing

<u>Do not let melt, stir until core temp reaches 1100°F</u>

3) Dump roasted ore on to metal surface and let cool

4) Pre-heat furnace to 1984° F

5) Thoroughly Pre-mix flux and Concentrate in separate mixing device before adding to the pre-heated furnace, (Do not mix in the coke yet).

6) Add the ore mixed with flux and copper to the furnace

7) Once the furnace is loaded, place the coke on top.

8) Increase temperature to 2250° F for 3 hours

9) Stir using a carbon rod every 45 minutes

10) After 3 hours in the furnace pour molten ore and copper into a mold, cool and separate out slag.

Basic Smelting with Lead

Note: Smelting with lead is basically the same process as the fire assay only on a larger scale.

Lead is very good at collecting precious metals and destroying the base metals leaving only the precious metals after oxidation coupled with very high confidence that all has been collected leaving nothing behind.

Soda Ash

Anhydrous Borax

Soda Ash

Household whole grain wheat flour

Litharge (Lead Oxide)

Potassium Nitrate (If a lot of Iron is present)

Flux components:

The following may be adjusted accordingly:

- 16 oz / 10lbs ore crushed to -20 Tyler mesh (or finer)
- 14.88 oz/ 9.3 lbs Soda Ash
- 1.6 oz/ 1 lb -Borax
- .64 oz/ 6.4 oz Household Whole Wheat Flour
- 14.88 oz/ 9.3 lbs , Litharge
- .64 oz/ 6.4 oz Silica
- 2250F for 1.2 hrs

Melting Verification procedure—

- Proper capacity Crucible furnace (Electric or Gas)
- Proper capacity Crucible
- Mold

Samples will need to be crushed to 20 Tyler mesh

Below are instructions for melting 1lb of Standard concentrates with flux

1. Have samples crushed to -20 mesh or finer and prepared for processing
2. Thoroughly Pre-mix flux and Concentrate in separate mixing device before adding to the pre-heated furnace.
3. Add the ore mixed with flux and litharge to the furnace.
4. Furnace at temperature to 2250° F for 1.2 hours
5. After 1.2 hours in the furnace pour molten ore and lead into a mold, cool and separate out slag

Cupellation must occur after this.

1. To Finish the process of the Lead Smelting process place the Lead Prill in a properly sized cupel to hold the mass once it is in its molten state.
2. Set Oven, (You must use a still oven no air turbulence), at 1850° F slightly crack door open (enough to allow oxygen but not enough to loose temperature).
3. Allow cupel enough time to absorb the oxidized lead until mass is solid
4. The material that is left should be pure precious metal upon completion.

Chapter 6: Assay by Fire (Fire Assaying)

Basic introduction to Assaying by Fire (Tools and Materials)

THE SMALL MINERS HELPER
A Basic overview of Mining and Processing in the US

All of the Basic tools that are needed to do an assay are critical since accuracy is a necessity and the fact that an assayer will work with very extreme temperatures and conditions. Therefore, contrary to what seems to be posted on the internet and some not so professional companies which offer assaying advice, are no shortcuts or cheap tools and materials.

One must remember when procuring products for assaying that one will find a wide array of producers, most manufacturers of these products are not located in the US, so when one wants to find a product from a distributor, you must know, test, or be familiar with the particular product before a decision is made to use this in production.

To start with, described below you will find some of the basic tools needed for assaying, note many are handmade since many of these tools have a tendency to be tailored for a particular application, furnace type, oven type etc.

- **Furnace:** (This can be either gas or electric), since the fuming from fluxes and metal vapors at high temperatures are so aggressive and volatile they have a tendency to attack the elements in an electric furnace, which can be very costly both in downtime and parts. Therefore it is recommended to use a gas fired furnace for reducing the sample mixture to a molten mass.
- **Assay oven or Kiln:** (Again this can be either Gas or electric), However the cupellation process must be done in a still air condition, with no drafts, therefore it is recommended for the small operation that this oven be electric.
- **Hand Tools / Hot Tools:** Most tools for the hot part of the assaying application are handmade they consist of tongs loaders and spatulas of many shapes, configurations, and sizes. This depending on such things as Top loading furnaces, Front loading Furnaces and ovens, distance needed between the assayer and hot work as well as controllability. Because of the strange configurations of these devices and lack of demand for them is the reason they are mostly handmade. A Pyrometer with a type 'K' Thermocouple should also be used to check temperatures if one is unsure about the temperature of the furnace
- **Miniature and Medical instruments:** An assayer will need such things as tweezers of different shapes and sizes, small enough to remove and manipulate some beads which can only be seen thru aid of magnification. Mostly by preference many of these can be found as Medical and Dental supplies. Magnifiers, Loupes, and a good microscope is a must. Optical comparators (Miniature), can also be used for determining an approximate bead weight thru size, using a mass scale.
- **Scales and accessories:** Generally a small operation must have 3 scales, The first must be capable of weighing larger masses up to 50Kg or 110 Lbs. These scales will generally cove receipt weight of any sample for testing. The second being a scale that will be used more than any other. This scale will be used in the range of 0 to 50 grams and measure in 10ths with a high degree of accuracy, used for measuring pulps and reagents. The 3rd or next Scale, highly necessary is the 1mg to 500g scale this scale is slightly expensive

depending on a new or used purchase and since it is designed for analytics and has no other purpose along with dependency on it being highly accurate. One must remember when dealing with all of the scales that it is highly necessary to calibrate all scales and the purchase and use of calibration weights are a must.

- **Sample splitter:** A sample splitter is an instrument which divides a primary sample into equal parts indicating a representative sample of the whole mass, the more times split the more accurate the representation. (Note there may still be an anomaly called nugget effect, in which a small particle of high grade material or metal may be in one pulp sample making the assay inaccurate, sometimes common in free milling ore, and can be eliminated by multiple samples of pulp being assayed).

- **Crushers and Pulverizers;** These are tools or machines necessary for comminution of the samples in preparation of pulps for an assay, it is desirable to have all material broken down into a pulp of less than 100-200M Tyler (149-74 microns) allowing good liberation and fracturing of the material so that it is more susceptible to particle collection during the Furnacing process. This breakdown is also preferential to most assayers. To do this the following is recommended.

  1. **Small laboratory Jaw Crusher:** To break down larger material for further processing and to fit in the smaller mills reducing time of comminution.
  2. **Impact Mill or Disc pulverizer:** For dry materials
  3. **Laboratory Ball mill or Tumbler with full charge of balls:** For wet grinding

- **Optional Tools / Safety Gear:**
- **Classifier Screens:** An assayer may find it necessary to do a size by size analysis where it is necessary to see before or after comminution, where the most values lie. This is done by screening the materials, usually using specific ASTM Tyler mesh sizes to see what percentages of value are in a particular particle size, afterward doing multiple assays on those particular samples to determine processing methods. There are many methods to accomplish the above screening from hand screening to Ro-tap and other automatic screeners.

- **Safety Gear:** This such as protective fire retardant clothing, Face and Eye protection should always be worn by any individual performing any and all of the analysis of the Fire assay. Extreme temperatures capable of causing maiming and/or death are without doubt very probable without the proper personal protective equipment commonly referred to as PPE. Equipment such as respirators and breathing apparatuses are a necessity as metal fluxing and assay fumes are poisonous. Litharge, Lead oxide is considered a poison fatal if inhaled or ingested. Nitric acid fumes can also be considered very deadly. If you can find safety gear that you think will benefit you, buy it and use it, you cannot take too many precautions. (Examples of Gear demonstrated in class).

Procedures for an Fire Assay: Please note, One very large and important step is for the assayer to keep extremely good redundant records, the next is total quality assurance by establishing a written procedure, updating and amending it periodically as needed, next are listed by function the steps in performing and living with a confident assay. Also note and understand that an assay no matter how large or small is only a representative of a sample accepted into your care.

One must also understand that a good assay carries no bias, and/or outside influence. The assayer must report exactly and only the seen results, any less or more is unacceptable. First part of the assaying is commonly referred to as firing, it requires a fireclay crucible capable of holding the sample and all of the reagents without boil over.

Startup
1. Prepare furnace and kiln start them up, so they may be brought up to temperature
2. Gather all tools and safety equipment needed put them in their prospective positions for use.
3. Move into sample area prepare sample by
   - Weighing and Logging in description and exact weight of sample
   - Clean or make sure all tools used for this assay are clean and free of debris (This is done as to never contaminate an assay whereby it would compromise its integrity.
   - Comminution of sample preparing for splitting (Wet samples may need to be dried afterward.
   - Splitting of sample to obtain a minimum of 3 samples of size and weight for firing. An example may be (3) 29.166 grams, so 90 grams of material would be adequate for this assay.
   - Bringing the assay to a readiness by taking a clean weight boat and weighing the proper assay weight before placing it in the crucible, in the case of three do this 3 times. Now With a ceramic crayon mark and identify the crucibles with the identification of the sample.
   - General rule only, Use 3 times your pre-mixed reagents to each part of sample weight, mix thoroughly using a wooden or plastic mixing stick.
   - Place crucibles in at temperature furnace for a minimum of 65 minutes.

If using an unknown style and manufacturer of crucible be sure to charge in first afterward removing it and checking for cracks and holes. The reason that pre-mixed flux (Reagents) were mentioned is that there may be certain alterations that might have to be made to tailor the flux to the ore, using a standard charting method.

> Finishing Firing
> 1. Check that mass is still in the crucible (No bubbling or boiling, if it is leave it in until it stops).
> 2. Prepare conical mold by heating to insure there is no moisture present..
> 3. Open Furnace, Remove crucible and pour into mold, one continuous pour rolling crucible to prevent dripping, (If lead spits from mold discard assay and perform a re-do.
>   - After cooling remove lead prill from mold
>   - Knock off slag
>   - Pound lead into a square

Now you are ready for Cupellation!

Procedures for an Fire Assay (Cupellation): The next step of the assay is called cupellation or Cupelling the lead prill. Normal usage requires a concaved surface made of either Bone ash or Magnesite, both being made of a very porous material capable of absorbing the lead which will oxidize in this stage leaving only the precious metal bead, referred to as a Cupel. Cupellation at 1775°F takes approximately 5 minutes per gram of lead under good conditions.

> 1. Weigh Prill and record (This will let you know approximately when the cupellation is finished so as not to disturb the process).
> 2. Using a Scribe Identify the Unit or Cupel
> 3. Place the Squared lead prill in the cupel
> 4. Place the Cupel in the Kiln
> 5. Check on Cupellation about 5 minutes before due time to finish.
> 6. Remove Cupel from oven, allow cooling.
> 7. Remove bead and weigh recording weight.

- ❖ On a 29.166 gram assay, the bead weight in milligrams is equivalent ounces per ton of precious metal.
- ⬥ You are now ready to see how much of this bead is Gold and how much is Silver!

Identification and parting of a precious metal bead: There are 4 ways to identify amounts of gold and silver in a PM bead.

> a. X-Ray Spectrometry (Very costly) Accurate
> b. Inter coupled Plasma testing (ICP) Extremely Costly, Accurate
> c. Colorimetric Charting (Least costly, accuracy depends on individual doing analysis) non destructive

    d. Dissolution using a heated 10% Nitric acid Solution dissolving all the silver in the bead. (Most accurate but requires a weighable amount of gold to determine value. (Inexpensive)

Parting using a heated 10% Nitric acid solution: The following steps will direct you on use of Nitric, acid dissolving the Silver in the bead to determine the percentages of Au and Ag.

1. Prepare for use of a clean parting cup
2. Turn heat source on to a medium heat
3. Put on rubber gloves ,add parting solution to parting cup (filling cup half way) and add the bead with a set of tweezers
4. Using tweezers set parting cup on hot plate (will see steam after a few minutes)
5. Allow ample time for bead to stop boiling
6. Remove from heat make sure mixture is still pour off excess liquid.
7. Move (Wash) the remaining material with distilled water.
8. Wash again with household ammonia
9. Wash a 3rd time with distilled water
10. Wash into a scorifier dish place a propane mini torch heating the mixture to anneal it to its natural color of gold. Wait for cooling
11. Place into a clean weight boat, weigh and calculate percentage of precious metal bead.
12. Record results.

Parting takes roughly 5 minutes to complete

Remember Assay and Metal Fumes are poisonous use adequate ventilation and Personal protective equipment at all time when performing these procedures.

Chapter 7: Selling Precious Metals and Dore'

There are many out there stating 'I Buy Gold' but beware! I have seen many Placer miners sell their gold at 65% to 70% market value. What's market value you ask? Well it is spot price for pure .999 or 24K Gold and most refiners pay 97 to 99% for the pure recovered metal (Most Gold, Karat, (Jewelry), Dore (Combined and smelted Gold and Silver), and Placer, is not pure Placer can range from 65% to 98.5% pure and generally contains an amount of silica, base and other metals). Settlement or payment can usually be paid in pure metal, by check, bank wire, or banked.

That's like trading a dollar bill for some pocket change! You should be paid what the value is. Such refiners as NTR metals pay 98% for all recovered gold but don't pay for Silver content unless it's very high. You have to set up an account with them, but it's worth it, you won't be cheated. Others pay 99% and pay for Silver but charge a refiners fee or contamination fee for such thing as tellurium, arsenic, iron, lead, etc. which may be found in your dore.

However these are all questions to be asked before making a deal with your metal. Dore in the industry is referred to as the combined metals of Gold and Silver (Not exactly the dictionary meaning as it is stated in percentages.

Then....there's the horror stories about the miner that sent 5 ounces of gold to a refiner and never got paid, well I have heard this for years but the refiners have no reason to cheat you, they work on such a small margin that it would not affect their bottom line. The case has been that if you send them junk you receive nothing, make sure to assay or have assayed what you send before you send it.

There are several refiners which advertise on the internet but not limited to:

NTR Metals

DH Fell

Republic

Disclaimer:

My Name is Norman Kendall I am a mining and processing consultant, by presenting the following information to you, it is my intention to only assist and provide information with mining facts, opinions and processes practiced today by, many mills and myself.—Some of this information may contain some unintentional errors and is not meant to replace proven methods, laws or procedures—I am not a Lawyer---All stated laws should be thoroughly proven before followed by an attorney. All opinions are of my own and stated by my own personal beliefs, cited in this book as a first amendment right to the US constitution. Some material has been copied from the internet where no copyright was identified on the pages (Such as the government information on mining claims and some commons photos), this information may no longer be valid as it is subject to change without notice—Common practices and processing may change with technology and may include differences not stated or may not work on certain ores. It is the sole responsibility of the person reading this book to investigate the processes and instructions before using them. There may be typos included in these writings as well. I make no guarantees as to the safety, practice, or outcome of any which have been stated.

www.ingramcontent.com/pod-product-compliance
Lightning Source LLC
Chambersburg PA
CBHW021025180526
45163CB00005B/2124